INTERNATIONAL CENTRE FOR MECHANICAL SCIENCES

ZLATKO JANKOVIĆ
UNIVERSITY OF ZAGREB

SELECTED TOPICS AND APPLICATIONS OF TENSOR ANALYSIS

COURSE HELD AT THE DEPARTMENTS
FOR MECHANICS OF DEFORMABLE BODIES
AND FOR HYDRO-AND GASDYNAMICS
JUNE - JULY 1970

UDINE 1970

ISBN 978-3-211-81165-8   ISBN 978-3-7091-2892-3 (eBook)
DOI 10.1007/978-3-7091-2892-3

Copyright 1970 by Springer-Verlag Wien
Originally published by Springer-Verlag Vienna in 1970

These lecture notes are manuscript of a paper which
will be published in TENSOR, vol. 22 (1971) N° 2.

PREFACE

This series of seven lectures entitled "Selected topics in the tensor calculus" is a natural continuation of the lectures "A contribution to the vector and tensor analysis", held at the International Centre for Mechanical Sciences in 1969. The essential points of a general scheme of the vector and tensor calculus described there in detail are briefly reviewed here for further development of the scheme. The central role in this development is played by the transposition operators which establish the one-to-one correspondence between the bra (ket) and ket (bra) vector spaces. The properties and significance of these operators in the scheme are discussed in full detail. One of the most important conclusions drawn is that the whole calculus can be formed with the help of the notion of absolute differential and is based on the identical vanishing of the absolute differentials of the fundamental and transposition operators. An illustrative example is given to show the generality and efficiency of the proposed scheme.

In the author's opinion, the aim of the lecture notes is best realized by the manuscript of his paper "On the transposition operators in a general vector and tensor calculus scheme", to be published in Tensor N.S. $\underline{22}$ (1971), N.2 which contains a systematic *presentation of the above mentioned subject.*

The author expresses his deepest appreciation and gratitude to the Authorities of CISM, particularly to the Secretary General Prof. L. Sobrero and to the Rectors Prof. O. Onicescu and Prof. W. Olszak for giving him this excellent opportunity to present his recent results of the study of the vector and tensor calculus to such a prominent audience in such a high-level international scientific institution.

Z. Janković

Zagreb, July 1970.

On the Transposition Operators

in a

Generalised Vector and Tensor Calculus Scheme

In previous papers [1], [2] and [3] we described the basic features of a generalized scheme of the vector and tensor calculus. In the present paper we investigate in detail the properties and role of the transposition operators in this scheme. In Chapter 1 the fundamentals of the scheme are described, while in Chapter 2 the basic properties of the transposition operators are established. In Chapter 3 the consequences of the required identical vanishing of the absolute differentials of the fundamental operator and of the transposition operators are investigated and the coefficients of connection are expressed in terms of the fundamental and transposition operator components. In Chapter 4 we discuss the role of the transposition operators in a space, i. e. their significance for the identification of the displacement vector components with the coordinate differentials and for the determination of the basis vector transformation coefficients. In Chapter 5 an important illustrative particular case of the scheme is considered.

1. The Basic Assumptions of the Scheme.

The basic assumptions of the proposed generalized scheme of the vector and tensor algebra and the notation used were described in [1]. Here we briefly recall that for the vectors of the four n-dimensional vector spaces: $\underline{a} = a^i \underline{e}_i \in \underline{X}$ (bra-down); $a^< = a_i e^i \in X^<$ (bra-up); $\overline{a} = {}^i a\, {}_i e \in \overline{X}$ (ket-down); $\overline{a} = {}_i a\, {}^i e \in {}^>X$ (ket-up) vector spaces (e_i,... the basis vectors, a^i,... the vector components), two kinds of multiplication were introduced which are of fundamental importance in building up the scheme. The s c a l a r product between a bra and a ket vector results in a scalar and the d i r e c t product of vectors results in a tensor, i.e. in a vector in the product vector space. Both kinds of products should exhibit the following properties: 1) they should be distributive in factors, 2) the extraction of scalars from factors should be allowed. Moreover, we have to assume the basis vector scalar products explicitly

(1.1) $\quad e^i{}_j e = {}^i\delta_j, \quad e_i{}^j e = {}_i\delta^j, \quad e^{ij}e = {}^i g^j, \quad e_i{}_j e = {}_i g_j;$

$$i,j = 1,2\ldots n.$$

Scalar and Direct Product

Here we have used the usual Kronecker δ symbol; $_ig_j$ and $^ig^j$ are $2n^2$ arbitrary, but assumed known, scalar quantities.

With the values (1.1) the scalar products of a bra and a ket vector have these explicit forms

$$a^<_>b = a_i{}^ib, \quad a_<^>b = a^i{}_ib,$$
$$a^{<>}b = a_i{}^ig^j{}_jb, \quad a_{<>}b = a^i{}_ig_j{}^jb, \quad (1.2)$$

where the double index summation convention has been applied.

For the direct product (denoted by the sign $\otimes$, if necessary) the associative law has also been required. The product vector space of r factor vector spaces represents a vector space in which the basis vectors can be taken to be direct products of basis vectors of r factor vector spaces, e. g.

$$^ie \otimes e_k \otimes e_p \equiv {}^ie\, e_k e_p, \quad {}^ie \in {}^>X(1), \quad e_k \in X(2)_<, \quad e_p \in X(3)_<$$

are the basis vectors in the product vector space $^{1>}X_{<2<3} \equiv {}^>X(1) \otimes X(2)_< \otimes X(3)_<$. The objects of the product vector space, i. e. tensors have the rank determined by the number of factor vector spaces, while their type is determined by the kind and order of factor basis vectors.

Thus, e. g.

(1.3) $$^{>}A_{<<} = {}_iA^{jk}\,{}^ie\,e_j\,e_k$$

is a third-rank tensor of ket-up, bra-down, bra-down type, briefly of the type $\left(\begin{smallmatrix}1\\\bullet\end{smallmatrix}\middle|\begin{smallmatrix}\bullet&\bullet\\1&1\end{smallmatrix}\right)$.

By convention, in the direct product the ket basis vectors have to be put to the left of the bra basis vectors ; in this case the sign $\otimes$ of the direct product is superfluous because the "active" sides of the vectors, i. e. the "valences", are not turned one to the other and no confusion with the scalar product is possible.

In the direct product the valences belonging to the same factor vector spaces can be allowed to act one on another through the scalar multiplication following the scheme (1.1); i. e. they can be contracted in pairs (the contracted pair of valences is denoted by a sign). In this sense the scalar product is a contracted direct product of a bra and a ket vector, e. g.

(1.4) $$a^{<} \otimes {}_{>}b = a^{<}{}_{>}b = a_i\,{}^ib.$$

This ability of the related valences of the same factor vector spaces to act one on another through the scalar multiplication, i. e. through the "saturation", justifies the term "operator" for tensors (vectors).

The following second-rank tensors (i.e. linear operators) can be formed using the scalars of the scheme (1.1) as their components:

i d e n t i t y operators

$$_>E^< = {_ie}\,\delta^i_j\,e^j, \qquad {^>E_<} = {^ie}\,e_i, \qquad (1.5a)$$

v a l e n c e r a i s i n g and l o w e r i n g operators

$$^>g^< = {^>E_<}\otimes {_>E^<} = {^>E_<}\,{_>E^<} = {^ie}\,_ig_j\,e^j,$$
$$_>g_< = {_>E^<}\otimes {^>E_<} = {_>E^<}\,{^>E_<} = {_ie}\,{^ig^j}\,e_j. \qquad (1.5b)$$

The requirement

$$_>g_<\,{^>g^<} = {_>E^<}, \qquad {^>g^<}\,_>g_< = {^>E_<}, \qquad (1.6)$$

ensures a one-to-one correspondence between the vectors of the bra (ket) vector spaces, and it is possible to conceive the connected vectors, e. g. $a^<$ and $a_< = a^<\,_>g_<$, as two forms ("up" and "down") of the same vector. Through the requirement (1.6) the nonsingular valence raising and lowering operators completely determine each other, one becoming the inverse operator to the other. Accordingly, of $2n^2$ components (1.1) of these operators, only n^2 components of one of them are at our disposal.

An important consequence of the requirement (1.6) is that the four operators (1.5) represent the four forms of the second-rank f u n d a m e n t a l tensor (operator) F

$$(1.7) \quad {}_>F^< = {}_>E^<, \quad {}_>F_< = {}_>F^<{}_>g_< = {}_>F^{(<)} = {}_>g_<,$$
$$ {}^>F^< = {}^>g^<, \quad {}_>F^< = ({}_>)F^< = {}_>g^<, \quad {}^>F_< = {}^>F^<{}_>g_< = {}^>g^<{}_>g_< = {}^>E_<.$$

It should be pointed out that the fundamental operators of the factor vector spaces act in the same manner on the respective valences in the product vector space, e. g. for a third-rank tensor we have

$$(1.8) \quad {}_{1>}A^{<2}{}_{<3} = {}_>g(1)_<{}^>A({}_<{}^>g(2)^<)_{<3} = {}^{(1>)}A({}_{<2})_{<3}.$$

Another important consequence of (1.6) is the identity of the four scalar products (1.2)

$$(1.9) \quad a^<{}_>b = a_<{}^>b = a^<{}^>b = a_<{}_>b = \mathbf{ab}.$$

2. The Transposition Operators.

The second rank t r a n s p o s i t i o n tensors (i.e. linear operators) T realize the

Basic Properties

correspondences of the vectors of the bra (ket) and ket (bra) vector spaces ([1], Ch. 3). We explain it on an example (the valence with an asterisk acts following the scheme (1.1) and the conjugate components belonging to the other saturated valence have to be taken)

$$T_<^{<*}{}_>a = \bar{a}_< , \quad \bar{a}_<{}^{*>}{}_>T = (T_<^{<*}{}_>a)^{*>}{}_>T = {}_>\bar{\bar{a}} , \qquad (2.1a)$$

i. e.

$$T^i_{j}{}_i e^{j*}(^k_{k}a \,_ke) = T^i{}_k{}^k a^* \, e_i = \bar{a}^i \, e_i ,$$
$$(T^i{}_r{}^r a^*)e_i \,^{*j}e \,_k e \,_j T = T^{i*}{}_r{}^r a \,_i T \,_k e = {}^k\bar{\bar{a}} \,_k e . \qquad (2.1b)$$

The requirement that the associated vector ${}_>\bar{\bar{a}}$ of the associated vector $\bar{a}_<$ should be identical with the original one ${}_>\bar{\bar{a}} \equiv {}_>a$ realizes a one-to-one correspondence characterized by

$$T_<^{<*} \otimes {}^{*>}{}_>T = {}_>E^< \quad \text{or} \quad T^{i*}{}_{ri}{}^k T = {}^k\delta_r . \qquad (2.2)$$

The four forms of the transposition operators obtained with the help of the fundamental operator (1.5) are as follows

$$T_<^{<*} , \qquad {}^{*>}{}_>T ,$$
$$T_<{}_{<*} = T_<(^{<*}) , \qquad {}_{*>}{}_>T = (^{*>})_>T , \qquad (2.3)$$

Chap. 2. The Transposition Operators

$$(2.3) \quad T^{<\,<*} = T(_<)^{<*}, \qquad T = ^{*>}(_>)T,$$
$$T^{<}{}_{<*} = T(_<)(^{<*}), \qquad {}_{*>}{}^{>}T = (^{*>})(_>)T.$$

Taking into account all possible correspondences of the described type (2.1) between the bra (ket) and ket (bra) vector spaces, eight relations are obtained in a manner analogous to that applied in deriving the relation (2.2)

$$(2.4) \quad \begin{aligned}
&T_{\bar{<}}{}^{<*} \otimes {}^{*\bar{>}}{}_{>}T = {}_{>}E^{<}, & &T_{<}{}^{\bar{<}*} \otimes {}^{*>}{}_{\bar{>}}T = {}^{>}E_{<}, \\
&T^{\bar{<}\,<*} \otimes {}_{*\bar{>}>}T = {}_{>}E^{<}, & &T_{\bar{<}<*} \otimes {}^{*>\bar{>}}T = {}^{>}E_{<}, \\
&T^{\bar{<}}{}_{<*} \otimes {}_{*\bar{>}}{}^{>}T = {}^{>}E_{<}, & &T^{<}{}_{\bar{<}*} \otimes {}_{*>}{}^{\bar{>}}T = {}_{>}E^{<}, \\
&T_{\bar{<}<*} \otimes {}^{*\bar{>}\,>}T = {}^{>}E_{<}, & &T^{<\,\bar{<}*} \otimes {}_{*>\bar{>}}T = {}_{>}E^{<}.
\end{aligned}$$

We shall denote the described one-to-one correspondence as case b). It is characteristic of this case that the components of the associated vector are expressed by the conjugate components of the original vector, as a result of applying the trans‍position operators with a valence with an asterisk (2.3). Case a) will denote the one-to-one correspondence where the conjugation does not occur, i.e. the transposition operators $T_{<}^{<}$,... in case a) leave the

Basic Relations 13

components of the original vector unchanged, the relations for these operators being of the same form as (2.3) and (2.4), but without an asterisk.

Because of the relations (2.3) and (2.4), in both cases only n^2 components of a determined form of the nonsingular transposition operator are at our disposal. Thus, for example, from the n^2 components $T^i_{\ k}$ of the nonsingular transposition operator $T_<^{<*}$, $\det(T^i_{\ k}) \neq 0$, the components $_i^rT$ of $^{*>}_{\ >}T$ are determined by equations (2.2), the components of all other forms (2.3) being subsequently determined with the help of fundamental operator (1.5).

The described one-to-one correspondences a) and b) between the bra (ket) and ket (bra) vector spaces with the help of the transposition operators (2.3) allow to take the connected vectors as two forms ("bra" and "ket") of the same vector, e. g. $_>a$ and $a_< = T_<^{<*}{}_>a$ for case b) or $_>a$ and $a_< = T_<^{<}{}_>a$ for case a).

An important consequence of introducing the transposition operators is the possibility of expressing the scalar product (1.9) by only one kind of vector components

$$\mathbf{a}\,\mathbf{b} = a^<{}_>b = a^<\,b^<{}_{*>}{}_>T = T^{<\,<*}{}_>a{}_>b =$$
$$= a_<{}^>b = a_<\,b_<{}^{*>\,>}T = T_{<\,<*}{}^>a{}^>b \ . \quad (2.5)$$

Now, from expressions (2.5) we easily find that for the commutative vector components the following assumed properties of the scalar product regarding the order of factor vectors, must reflect themselves in the analogous properties of the transposition operators

(2.6)

	$\mathbf{a}\,\mathbf{b} =$	$T_{ij},\ T^{ij},\ _{ij}T,\ ^{ij}T$
case a)	$\mathbf{b}\,\mathbf{a}$, symmetric	symmetric
	$-\mathbf{b}\,\mathbf{a}$, antisymmetric	antisymmetric
case b)	$(\mathbf{b}\,\mathbf{a})^*$ Hermitian	Hermitian
	$-(\mathbf{b}\,\mathbf{a})^*$ antihermitian	antihermitian.

It is easily seen that the transposition operators in the product vector space are composed as direct products of factor space transposition operators. Their properties and relations are obvious generalizations of those discussed here, as will be shown in a separate paper.

In the papers [1] and [2] the simplest transposition operators were introduced and their properties and consequences investigated. These are cases A and B: (case b), for case a) omitt the asterisks) :

Simplest Cases

		case A assump. conseq.		case B assump. conseq.	
$T_{<}{}^{<*}$	$T^i{}_j$	$\delta^i{}_j$			${}_ig^*_j = {}^jg^i$
$T_{<}{}_{<*}$	T^{ij}		${}^ig^{j*} = {}^jg^i$	δ^{ij}	
$T^{<}{}^{<*}$	T_{ij}		${}_jg_i = {}_ig^*_j$	δ_{ij}	
$T^{<}{}_{<*}$	$T_i{}^j$	$\delta_i{}^j$			${}_jg_i = {}^ig^{j*}$
${}_{*>}{}^{>}T$	${}^j{}_iT$	${}^j{}_i\delta$			${}^jg^{i*} = {}_ig_j$
${}_{*>}{}_{>}T$	${}_{ji}T$		${}_jg^*_i = {}_ig_j$	${}_{ji}\delta$	
${}^{*>}{}_{>}T$	${}^{ji}T$		${}^ig^j = {}^jg^{i*}$	${}^{ji}\delta$	
${}^{*>}{}^{>}T$	${}_j{}^iT$	${}_j{}^i\delta$			${}^ig^j = {}_jg^*_i$

(2.7)

Hence we conclude that the valence raising and lowering operators should have the properties ([1], Ch. 3)

case	A	B
a	symmetric	orthogonal
b	Hermitian	unitary.

(2.8)

We add the following remark to table (2.7). As we saw earlier (cf. (2.3) and (2.4)) n^2 components of a form of transposition operator determine the components of all other forms of transposition operators (2.3) with the help of the fundamental operator (1.7). In case A (B) (2.7) we assumed values of $2n^2$ components of two different forms of transposition operators. The consequence is the reduction of freedom in the choice of n^2 components (i. e. of their $2n^2$ real and imaginary parts), of the valence raising (or lowering) operators. This manifests itself in the indicated properties of these operators in table (2.8). Explicitly, the number of parts of complex components of valence raising (or lowering) operator at our disposal is ([1] , Ch. 3) : Aa) $n(n+1)$, Ab) n^2, Ba) $n(n-1)$, Bb) n^2.

3. The Absolute Differentials of the Transposition Operators.

The systematic formation of the vector and tensor analysis based on the notion of the "parallel displacement" of the field quantities was extensively discussed in ref. [2] . This specific one-to-one correspondence establishes an isomorphism between the corresponding n-dimensional vector spaces attached to the neighbouring points $P(x^k)$ and $Q(x^k+dx^k)$ of an m-dimensional manifold, x^k being assumed to be continuously varying real parameters (coordinates).

Absolute Differential

For the parallelly displaced field quantities the following properties are required

$$[A(P) \pm B(P)]_Q = A(P)_Q \pm B(P)_Q,$$
$$[A(P) \cdot C(P)]_Q = A(P)_Q \cdot C(P)_Q, \qquad (3.1)$$

where $A(P)_Q \in X(Q)$ is written for the parallelly displaced field quantity corresponding to the field quantity $A(P) \in X(P)$. The dot in the second relation (3.1) may represent each of the possible kinds of multiplication which can occur in the scheme, i. e. the multiplication by a scalar, a scalar product or a direct product. The first order part (in parameter differentials) of the difference between the field quantity $A(Q)$ and the parallelly displaced quantity $A(P)_Q$ is called the a b s o l u t e differential

$$\Delta A(Q) = dx^k \nabla_k A(Q) = A(Q) - A(P)_Q, \qquad (3.2)$$

∇_k being called the k-th absolute derivative. The rules for the absolute differential (derivative) of a sum (difference) or a product of the field quantities are an immediate consequence of the requirements (3.1) and (3.2)

$$\Delta(A \pm B) = \Delta A \pm \Delta B, \quad \Delta(A \cdot C) = (\Delta A) \cdot C + A \cdot (\Delta C). \quad (3.3)$$

Thus, to determine the parallelly dis-

Chap. 3. The Absolute Diff. of the Transposition Op.

placed field quantity (to the first order in parameter differentials), it is necessary to determine its absolute differential. It is therefore necessary to assume explicitly

1) the absolute differentials of the corresponding basis vectors ([2], (1.3)c))

(3.4)
$$\Delta^i e = dx^k {}^i(_j\Gamma)_k {}^j e \ , \quad \Delta_i e = dx^k {}_i({}^j\Gamma)_{kj} e \ ,$$
$$\Delta e^i = dx^k (\Gamma_j)^i_k e^j \ , \quad \Delta e_i = dx^k (\Gamma^j)_{ki} e_j \ .$$

2) that the scalar quantity be unaffected by the parallel displacement ([2], (1,3d)) or, equivalently, that the absolute and ordinary differentials of a scalar quantity are identical

(3.5)
$$\Delta f = df \ .$$

The four sets of mn^2 functions of parameters Γ appearing in the relations (3.4) (index k running from 1 to m, indices i and j from 1 to n) are called the coefficients of connection, because they realize the connection of the vector spaces $X(P)$ and $X(Q)$.

From the scheme (1.1) we easily deduce the extremely important result that the absolute differentials of all the four forms of the fundamental tensor field (1.8) disappear identically

$$(\Gamma_i)_k^{\dot{j}} + {}_i({}^{\dot{j}}\Gamma)_k = 0, \qquad \Delta_{>}F^{<} = \Delta_{>}E^{<} \equiv 0;$$

$$(\Gamma^{\dot{i}})_{k\dot{j}} + {}^{\dot{i}}({}_{\dot{j}}\Gamma)_k = 0, \qquad \Delta^{>}F_{<} = \Delta^{>}E_{<} \equiv 0;$$

$$\partial_k {}^{\dot{i}}g^{\dot{j}} = (\Gamma_r)_k^{\dot{i}} \, {}^r g^{\dot{j}} + {}^{\dot{i}}g^r {}^{\dot{j}}({}_r\Gamma)_k, \qquad \Delta_{>}F_{<} = \Delta_{>}g_{<} \equiv 0;$$

$$\partial_k {}_{\dot{i}}g_{\dot{j}} = (\Gamma^r)_{k\dot{i}} \, {}_r g_{\dot{j}} + {}_{\dot{i}}g_r {}_{\dot{j}}({}^r\Gamma)_k, \qquad \Delta^{>}F^{<} = \Delta^{>}g^{<} \equiv 0.$$

(3.6)

The first column follows immediately from (1.1) by applying the rules (3.3) and the expressions (3.4) and (3.5). The second column is obtained from the first one by applying the rules (3.3) to the expressions (1.7). Thus, the fundamental tensor field has the property

$$F(P)_Q = F(Q). \qquad (3.7a)$$

This means that the isomorphism of the bra (ket) vector spaces realized by the fundamental operator is conserved in the parallel displacement or, equivalently, the fundamental operator and the absolute differential commute, e. g.

$$\Delta^{>}a = \Delta({}^{>}g^{<}{}_{>}a) = {}^{>}g^{<}\Delta_{>}a.$$

Now, we make an analogous requirement that the isomorphism between the bra (ket) and ket (bra) vector spaces realized by the transposition operator field should be conserved in the parallel dis-

Chap. 3. The Absolute Diff. of the Transposition Op.

placement

(3.7b) $$T(P)_Q \equiv T(Q),$$

or equivalently, the transposition operators and the absolute differential should commute, e. g.

$$\Delta\, a_< = \Delta(T_<{}^{<*}\,{}_>a) = T_<{}^{<*}\, \Delta_> a\,,$$

i. e.

(3.8) $$\Delta\, T_<{}^{<*} \equiv 0 \quad \text{or} \quad \left(T(P)_<{}^{<*}\right)_Q \equiv T(Q)_<{}^{<*}.$$

It is sufficient to require that the absolute differential of only one form of the transposition operator fields (2.3) should vanish identically. Because of the relations (2.4) the absolute differentials of all other forms have to vanish identically, too. Thus, e. g. from the assumption (3.8) the relations (2.4) and (3.3) yield

(3.9a) $$\Delta(T_<{}^{<*} \otimes {}^{*>}_>T) = T_<{}^{<*} \otimes \Delta\, {}^{*>}_>T \equiv 0\,.$$

By taking the direct product with ${}^{*>}_>T$ and contracting upon the double dashed valence, the relation (3.9a) takes the form

(3.9b) $$T_<{}^{\overline{<*}} \otimes \Delta\, {}^{*>}_>T \otimes {}^{*>}_{\overline{>}}T = {}^>E_< \otimes \Delta\, {}^{*>}_>T = \Delta\, {}^{*>}_>T \equiv 0,$$

where the relations (2.4) have been taken into account again. The identical vanishing of the other six forms of the transposition operators (2.4) is a consequence of the relations (3.8) and (3.9) and of the identical vanishing of the absolute differentials of the fundamental operator (3.6), e. g.

$$\Delta T_{<\,<*} = \Delta(T_<^{<*}, g_<) = \Delta(T_<^{<*}), g_< = 0. \quad (3.9c)$$

In the case of real parameters we easily prove that for the valences with an asterisk the relations of the form (3.4) containing conjugate complex coefficients of connection and basis vectors with an asterisk on the right-hand side are valid, e. g. from

$$\Delta(e_{i\,*}, {}^>a) = \Delta_i a^* \quad (3.10a)$$

with the help of the relations (3.3) and (3.4) (the relation (3.10 a) being valid for every vector ${}^>a$) it follows

$$\Delta e_{i\,*} = dx^k (\Gamma^j)^*_{ki} e_{j\,*} . \quad (3.10b)$$

Thus, the absolute differentials of all the eight forms of the transposition operators (2.3) have to vanish identically. For them we compose a table analogous to that for the fundamental operator (3.6)

Chap. 3. The Absolute Diff. of the Transposition Op.

(3.11)
$$\Delta T_{<}{}^{<*} \equiv 0, \quad \partial_k T^p{}_i + T^q{}_i (\Gamma^p)_{kq} + T^p{}_q (\Gamma_i)^{q*}{}_k = 0,$$

$$\Delta T_{<\,<*} \equiv 0, \quad \partial_k T^{pi} + T^{qi} (\Gamma^p)_{kq} + T^{pq} (\Gamma^i)^{*}{}_{kq} = 0,$$

$$\Delta T^{<\,<*} \equiv 0, \quad \partial_k T_{pi} + T_{qi} (\Gamma_p)^q{}_k + T_{pq} (\Gamma_i)^{q*}{}_k = 0,$$

$$\Delta T^{<}{}_{<*} \equiv 0, \quad \partial_k T_p{}^i + T_q{}^i (\Gamma_p)^q{}_k + T_p{}^q (\Gamma^i)^{*}{}_{kq} = 0,$$

$$\Delta^{*>}{}_{>}T \equiv 0, \quad \partial_k {}^p{}_i T + {}^q{}_i T ({}^p\Gamma)_k + {}^p{}_q T {}^q({}_i\Gamma)^{*}{}_k = 0,$$

$$\Delta_{*>\,>}T \equiv 0, \quad \partial_k {}^{ip}T + {}^{iq}T ({}^p\Gamma)_k + {}^{qp}T ({}^i\Gamma)^{*}{}_k = 0,$$

$$\Delta^{*>\,>}T \equiv 0, \quad \partial_k {}_{ip}T + {}_{iq}T {}^q({}_p\Gamma)_k + {}_{qp}T {}^q({}_i\Gamma)^{*}{}_k = 0,$$

$$\Delta_{*>}{}^{>}T \equiv 0, \quad \partial_k {}^i{}_p T + {}^i{}_q T {}^q({}_p\Gamma)_k + {}^q{}_p T {}^q({}^i\Gamma)^{*}{}_k = 0.$$

From the results (3.11) valid for case **b)** analogous results are obtained for case **a)** by omitting the asterisk in (3.11).

By inspection of the relations (3.6) we see that the first two reduce the number of $4mn^2$ coefficients of connection (introduced into the scheme through (3.4)), to $2mn^2$ coefficients, e.g. $(\Gamma^r)_{\iota i}$ and $_i(^r\Gamma)_k$. Furthermore, it is easy to conclude that the last two sets of relations (3.6) are mutually dependent because of (1.6), only one set of them being independent.

Coefficients of Connection

This independent set contains only mn^2 relations. Therefore mn^2 coefficients of connection are left undetermined by the identical vanishing of the absolute differentials of the fundamental operator (3.6).

The required identical vanishing of the absolute differentials of the transposition operators (3.11) supplies just one set of mn^2 independent relations for the coefficients of connection, since of the eight sets of relations (3.11), only one is independent. Thus, in general, the identical vanishing of the absolute differentials of the fundamental operator field (3.6) and of the transposition operator field (3.11) completely determine the coefficients of connection.

To prove this, we take the last relation (3.6) and the fifth relation (3.11)

$$\Delta^{>}g^{<} \equiv 0, \qquad \partial_k {}_i g_j = (\Gamma^r)_{ki\,r} g_j + {}_i g_r {}_j (^r\Gamma)_k ,$$
$$\Delta^{*>}{}_{>}T \equiv 0, \qquad \partial_k {}_i{}^p T = (\Gamma^r)^*_{ki\,r}{}^p T - {}_i{}^r T {}_r (^p\Gamma)_k . \qquad (3.12)$$

Now, we multiply the first relation by ${}^j g^q ({}^p g^i)$ and the conjugate second relation by $T^q{}_p$ (the second relation by $T^{i*}{}_j$). Then, taking into account the basic relations for the fundamental operator (1.6) and for the transposition operators (2.4), we obtain by subtraction (addition) these two relations

$$\partial_k {}_i g_j {}^j g^q - T^q{}_p \partial_k {}_i{}^p T^* = {}_j (^r\Gamma)_k {}_i g_r {}^j g^q + {}_j (^r\Gamma)^*_k T^q{}_r {}_i{}^j T^* , \qquad (3.13)$$

$$(3.13) \quad {}^P g^i{}_k \partial_i g_j + T^{i*}_{j} \partial_{ki} {}^P T = (\Gamma^r)_{ki} {}^P g^i{}_r g_j + (\Gamma^r)^*_{ki} T^{i*}_{j} {}_r {}^P T .$$

By taking separately the real and imaginary parts of the first and second relation (3.13) we obtain $2mn^2$ equations for $2mn^2$ real and imaginary parts of $({}^r\Gamma)_{jk}$ and $(\Gamma^r)_{ki}$, respectively ; hence these coefficients can be determined in terms of the fundamental and transposition operator components.

4. The Transposition Operators in a Metric Space.

In ref. [3] the case $m = n$, i. e. the case of equal dimensions of the vector spaces and of the parameter manifold was discussed. Such a manifold with a structure determined by the fundamental tensor and by the coefficients of connection may be called a space. We introduced the displacement vector **dx** which should connect the "distance" of the points $P(x^k)$ and $Q(x^k + dx^k)$ with the coordinate (parameter) differentials in the chosen system of real coordinate lines. Because of the relations (1.5) and (2.3) the four forms of the displacement vector can be written with only one kind of components $(dx)^i$

Displacement Vector

$(dx)_<=$	$(dx)^< =$	$_>(dx) =$	$^>(dx) =$
$(dx)^i e_i$	$(dx)_i e^i$	$^i(dx) \, _i e$	$_i(dx) \, ^i e$
	$(dx)^j \, _j g_i \, e^i$	$(dx)^j \, _j^i T \, _i e$	$(dx)^j \, _{ji}^{\,i} T \, ^i e$

(4.1)

The scalar product of the displacement vector by itself – the square of the norm of the displacement vector – should by assumption be equal to the square of distance $\overline{PQ} = ds$ of the points $P(x^k)$ and $Q(x^k + dx^k)$, i. e.

$$\overline{PQ}^2 = ds^2 = (dx)_< \, ^>(dx) = (dx)_< (dx)_< \, ^{*><>}T =$$
$$= (dx)^i (dx)^k \, _{ki}T \, .$$

(4.2)

Now, we require the bra-up components of the displacement vector $(dx)^i$ to be identical with the coordinate (parameter) differentials in every coordinate system

$$(dx)^i \equiv dx^i, \qquad i = 1, 2, \ldots, n \, .$$

(4.3)

Consequently, we have

$$(dx)_< = dx^i \, e_i \, , \quad (dx)^< = dx^j \, _j g_i \, e^i \, , \quad _>(dx) = dx^j \, _j^i T \, _i e \, ,$$
$$^>(dx) = dx^j \, _{ji}^{\,i} T \, ^i e \, , \quad ds^2 = dx^i \, dx^k \, _{ki}T \, .$$

(4.4)

Thus, by the identification (4.3) the corresponding basis vectors ([2], Intr.) are connected with the coordinate lines in a specific way. We call them the tangent basis vectors, because for the points P and Q on a coordinate line the displacement vector and the respective bra-down basis vectors are proportional (e. g. $(dx)_{1<} = dx^1 e_1$). Further, by the identification (4.3) the transposition tensor $^{*><}T$ acquires a specific role, i. e. it becomes the m e t r i c tensor of the space. Its importance for the formation of the expression for the distance $\overline{PQ}$ is evident from the expression (4.4).

The identification (4.3) which should be valid for every coordinate system has the following consequences for the determination of the transformation coefficients ([1], Ch. 4) which describe the transition from one system of tangent basis vectors to another system of tangent basis vectors, i.e. from one system of coordinate lines x^k, $k = 1,...,n$, to an other system of coordinate lines $x^{\prime\dot{j}}$, $\dot{j} = 1,...,n$. Thus, for the two systems of coordinate lines

(4.5) $\quad x^k = x^k(x^{\prime\dot{j}})$, $x^{\prime\dot{j}} = x^{\prime\dot{j}}(x^k)$; x^k, $x^{\prime\dot{j}} \in C^N$,

we should have the following identities in the coordinate differentials for the displacement vector **dx** con-

Transformation Coefficients

necting the points $P(x^k) \equiv P(x'^j)$ and $Q(x^k + dx^k) \equiv Q(x'^j + dx'^j)$

$$(dx)_< = (dx)_{<'} = (dx)_< {}^>E'_< = (dx)_{<'} {}^>E_< , \qquad (4.6a)$$

i. e.

$$(dx)_< = e_k dx^k = e'_j dx'^j = (e_k {}^j e') dx^k e'_j = (e'_j {}^k e) dx'^j e_k =$$
$$= e_k (\partial x^k/\partial x'^j) dx'^j = e'_j (\partial x'^j/\partial x^k) dx^k . \qquad (4.6b)$$

By a comparison of these expressions (4.6 b) we conclude that the transformation coefficients are equal to

$$(e_k {}^j e') = \partial x'^j/\partial x^k , \quad (e'_j {}^k e) = \partial x^k/\partial x'^j . \qquad (4.7)$$

By a completely analogous treatment of the other forms of the displacement vector (4.1) with the help of formulas (1.5) and (2.3) we obtain the expressions for the transformation coefficients

$$(dx)^< , \quad (e^k {}_j e') = (\partial x'^q/\partial x^p) {}_q g'_j {}^k g^p ,$$

$$(e'^j {}_k e) = (\partial x^p/\partial x'^q) {}_p g_k {}^j g'^q ,$$

$$_>(dx) , \quad (e'^j {}_k e) = (\partial x'^q/\partial x^p) T^{p*}_{kq} {}^j T' , \qquad (4.8)$$

$$(e^k {}_j e') = (\partial x^p/\partial x'^q) T'^{q*}_{jp} {}^k T ,$$

28 Chap. 4. The Transposition Oper. in a Metric Space

$$(4.8) \quad {}^{>}(dx) \, , \quad (e_{\dot{j}}^{\,k} e) = (\partial x'^{k}/\partial x^{p}) T^{pk*}{}_{q\dot{j}} T' \, ,$$

$$(e_{k}{}^{\dot{j}} e') = (\partial x^{p}/\partial x'^{q}) T'^{q\dot{j}*}{}_{pk} T \, .$$

Now, if the fundamental tensor and the transposition tensors are explicitly given in every coordinate system, all transformation coefficients (4.8) are determined by these tensors and the partial derivatives of the transformation functions (4.5). Inversely, knowing the transformation of the coordinate lines, $2n^2$ coefficients of transformation are determined by (4.7). The knowledge of additional $2n^2$ coefficients of transformation $(e^{k}{}_{\dot{j}} e')$ and $(e'^{\dot{j}}{}_{k} e)$ independent of the former ones (4.7), is necessary to determine the fundamental tensor and the transposition tensors in the transformed coordinate system from the known fundamental tensor and transposition tensor in the starting coordinate system. In this case, from the relations (4.8), (1.6) and (2.4) we obtain the following expressions for the components of the fundamental and transposition tensors

$$(4.9) \quad {}_{q}g'_{\dot{j}} = (\partial x^{p}/\partial x'^{q})\, {}_{p}g_{k}(e^{k}{}_{\dot{j}} e') \, , \quad {}^{\dot{j}}g'^{q} = (\partial x'^{q}/\partial x^{p})\, {}^{k}g^{p}(e'^{\dot{j}}{}_{k} e) \, ,$$

$$\quad {}_{q}{}^{\dot{j}}T' = (\partial x^{p}/\partial x'^{q})\, {}^{k}_{p}T(e'^{\dot{j}}{}_{k} e) \, , \quad T'^{q*}{}_{\dot{j}} = (\partial x'^{q}/\partial x^{p})\, T^{p*}{}_{k}(e^{k}{}_{\dot{j}} e') \, .$$

5. An Illustrative Particular Case.

To illustrate the generality of the described scheme (compare also ([1], Ch. 3), ([2], Ch.2)) we apply the obtained results to case A (2.7), the basis vectors being connected in the manner $e^i \leftrightarrow {}^i e$, $e_i \leftrightarrow {}_i e$, and the vector components

$$
\begin{array}{c|cccc}
 & a^i & {}^i a & a_i & {}_i a \\
\hline
a) & a^i & a^i & a_i & a_i \\
b) & a^i & a^{i*} & a_i & a_i^*
\end{array}
\qquad (5.1)
$$

The relations (3.6) for the fundamental tensor components are valid in their original form in case A, while the valence raising and lowering operators (1.5b) are symmetric in case Aa) and Hermitian in case Ab). Particularly, they are real symmetric tensors of the second rank for a real vector space. Because of the relations (2.7) and (3.6) the eight sets of relations (3.11) for the transposition operator components reduce to four sets

$$(\Gamma^i)_{kp} + (\Gamma_p)^{i*}_{\ k} = 0,$$
$$({}^p\Gamma)^*_{i\ k} + {}^p({}_i\Gamma)_k = 0, \qquad (5.2)$$

Chap. 5. An Illustrative Particular Case

$$(5.2) \quad \partial_k {}_i g_p + {}_i g_q {}^q ({}_p \Gamma)_k^* + {}_q g_p {}^q ({}_i \Gamma)_k = 0 ,$$

$$\partial_k {}^p g^{i*} + {}^q g^{i*} {}_q ({}^p \Gamma)_k^* + {}^p g^q {}_q ({}^i \Gamma)_k = 0 ,$$

where the immediate consequences of the first two relations (3.6)

$$(5.3) \quad {}^i ({}_p \Gamma)_k = (\Gamma_p)_k^{i*} ,$$

$$\quad \quad \quad {}_p ({}^i \Gamma)_k = (\Gamma^i)_{kp}^* ,$$

have been taken into account. The last two expressions (5.2) are identical with the last two expressions (3.6) because of the Hermitian property of the valence raising and lowering operators in case **Ab**). The relations for case **Aa**) follow immediately from formulas (5.2) and (5.3) by omitting the asterisks.

In case **Ab**), (**Aa**)), the relations (3.13) which connect the coefficients of connection $(\Gamma^r)_{ki}$ and ${}_j(\Gamma)_k$ with the fundamental and transposition tensor components, become indentical with the fourth set of relations (3.6), i.e. taking into account the formulas (5.2) and (5.3) the two relations reduce to the expression

$$\partial_k {}_i g_j = {}_j ({}_i \Gamma)_k + (\Gamma_j)_{ki} .$$

Because of the expressions (5.3), the

obtained expression (i.e. the fourth relation of the (3.6) in case Ab) can be written in the form

$$\partial_k {}_i g_j = (\Gamma_{\cdot})_{jki} + (\Gamma_{\cdot})^{*}_{ikj} . \qquad (5.4)$$

Since the fundamental tensor in case Ab) is Hermitian, we obtain from (5.4) the following expressions for the symmetric and antisymmetric parts (in the pair of indices i and j) of the coefficients of connection $(\Gamma_{\cdot})_{ikj}$

$$\partial_k {}_i g_j^{1S} = 2\mathrm{Re}(\Gamma_{\cdot}^{S})_{jki} , \quad \partial_k {}_i g_j^{2A} = 2\mathrm{Im}(\Gamma_{\cdot}^{A})_{jki} . \qquad (5.5)$$

The formula (5.4) represents only mn^2 independent relations (5.5) for $2mn^2$ real and imaginary, symmetric and antisymmetric, parts of the coefficients of connection. The consequence of this fact is that mn^2 parts $\mathrm{Re}(\Gamma_{\cdot}^{A})_{jki}$, $\mathrm{Im}(\Gamma_{\cdot}^{S})_{jki}$ remain undetermined by (5.4), i. e. they are at our disposal.

Since the fundamental tensor in case Aa) is symmetric, from the relation (5.4) we derive the expression

$$\partial_k {}_i g_j^{S} = 2(\Gamma_{\cdot}^{S})_{jki} , \qquad (5.6)$$

while the $mn(n-1)/2$ antisymmetric parts $(\Gamma_{\cdot}^{A})_{jki}$ have been left undetermined by (5.4), i. e. they are at our disposal.

Chap. 5. An Illustrative Particular Case

In the case of a space ([3], Ch. 2) n equals m and the coefficients of connection can also be analysed with respect to the symmetry in outer indices. In the particular case **Aa)** $(\Gamma.)_{ikj}^{A} = 0$ we find that for the symmetric (in outer indices) coefficients of connection the relation (5.4) takes the form

$$(5.7) \qquad \partial_{k\,i}g_{j}^{S} = (\Gamma.)_{jki}^{S} + (\Gamma.)_{ikj}^{S} .$$

Explicit expressions for the symmetric (in outer indices) coefficients of connection in terms of partial derivatives of the symmetric fundamental tensor components

$$(5.8) \qquad (\Gamma.)_{ikj}^{S} = \begin{Bmatrix} i \\ kj \end{Bmatrix}^{S} = \tfrac{1}{2}(\partial_{k\,i}g_{j}^{S} + \partial_{j\,i}g_{k}^{S} - \partial_{i\,k}g_{j}^{S}) ,$$

can be easily obtained by combining the relations (5.7) in the way indicated on the right-hand side.

In the particular case **Ab)** $\operatorname{Re}(\Gamma.)_{jki}^{A} \equiv 0$, $\operatorname{Im}(\Gamma.)_{jki}^{S} \equiv 0$ we obtain in a similar manner

$$(5.9) \qquad (\Gamma.)_{rkj} = \begin{Bmatrix} r \\ kj \end{Bmatrix}^{1S} + i \begin{Bmatrix} j \\ kr \end{Bmatrix}^{2A} .$$

In the real case **A** an arbitrary antisymmetric second-rank tensor ${}^{>}g^{A<} = {}^{i}e_{j\,i}g_{j}^{A} e^{j}$ can be used to express the antisymmetric part of the coefficients of connection at our disposal

Metric Tensor

$$2(\Gamma_{j}^{A})_{ki} = \partial_k \, {}_i g_{j}^{A} . \qquad (5.10)$$

Then, in terms of the b a s i c tensor with components

$$ {}_i \overset{\pm}{g}_j = {}_i g_j^S \pm {}_i g_j^A \qquad (5.11)$$

the coefficients of connection follow from the relations (5.6) and (5.10) in the form

$$2(\Gamma_{j})_{ki} = \partial_k \, {}_i \overset{+}{g}_j . \qquad (5.12)$$

The displacement vector in case A has the following four forms

$$(dx)_< = dx^i \, e_i \, , \quad (dx)^< = dx^j \, {}_j g_i \, e^i ,$$
$$_>(dx) = dx^i \, {}_i e \, , \quad {}^>(dx) = {}_i g_j \, dx^j \, {}^i e . \qquad (5.13)$$

The expression for the square of distance (4.4) for case A (2.7)

$$ds^2 = {}_i g_k \, dx^i \, dx^k = {}_i g_k^S \, dx^i \, dx^k \qquad (5.14)$$

shows that the symmetric part of the fundamental tensor determines the metrics of the space, i. e. is identical with the m e t r i c tensor, while the antisymmetric part (existing in case Ab)) does not affect the metrics at all.

The transformation coefficients (4.7)

and (4.8) in case A are completely determined by the partial derivatives of the transformation functions (4.5), if we require that the transposition operators be given by the expression (2.7, case A) in every tangent basis vector system. Then, from the last two formulas it follows

$$
(5.15) \quad \begin{aligned}
{}_q{}^{\dot{\jmath}}\delta &= (\partial x^p/\partial x'^q)\,(e'^{\dot{\jmath}}{}_k e)\,{}_p{}^k\delta \ , \\
\delta^q{}_{\dot{\jmath}} &= (\partial x'^q/\partial x^p)\,(e^k{}_{\dot{\jmath}} e')\,\delta^p{}_k \ , \\
(e'^{\dot{\jmath}}{}_k e) &= \partial x'^{\dot{\jmath}}/\partial x^k = (e_k{}^{\dot{\jmath}}e') \ , \\
(e^k{}_{\dot{\jmath}} e') &= \partial x^k/\partial x'^{\dot{\jmath}} = (e'_{\dot{\jmath}}{}^k e) \ ,
\end{aligned}
$$

in accordance with the fact that the scalar product (1.9) is Hermitian (symmetric) by (2.8) in case Ab) (Aa)) while the partial derivatives of the transformation functions (4.5) are real quantities for real coordinates.

We conclude with the remark that now the terms "up" and "down" could be exchanged with the terms "covariant" and "contravariant", respectively, when referring to valences (opposite terms have to be used for component indices). In accordance with the usual convention these terms can be used for valences and vector and tensor components which are transformed

Remarks

from the original system into the primed system of coordinate lines (or tangent basis vectors) with the transformation coefficients $\partial x^k/\partial x'^j$ and $\partial x'^j/\partial x^k$, respectively.

The real case **A**, extensively discussed in ref. [3], is only one special realization of the proposed generalized vector and tensor calculus scheme. The Riemannian space characterized by the symmetric real metric tensor and real symmetric (in outer indices) coefficients of connection is even a more particular realization of the real case **A**. From this fact the generality of the described vector and tensor calculus scheme should be evident.

References.

[1] Z. Janković : A contribution to the vector and tensor algebra. Tensor, N.S., 21 (1970) 151-166.

[2] Z. Janković : A contribution to the vector and tensor analysis I. Tensor, N.S., 21 (1970) 167-185.

[3] Z. Janković : A contribution to the vector and tensor analysis II. Tensor, N.S., 21 (1970) 189-203.

Contents.

	Page
Preface....................................	3
1. The Basic Assumptions of the Scheme.......	6
2. The Transposition Operators...............	10
3. The Absolute Differentials of the Transposition Operators...................	16
4. The Transposition Operators in a Metric Space............................	24
5. An Illustrative Particular Case..........	29
References.................................	37

MIX
Papier aus verantwortungsvollen Quellen
Paper from responsible sources
FSC® C105338

If you have any concerns about our products,
you can contact us on
ProductSafety@springernature.com

In case Publisher is established outside the EU,
the EU authorized representative is:
**Springer Nature Customer Service Center GmbH
Europaplatz 3, 69115 Heidelberg, Germany**

Printed by Libri Plureos GmbH
in Hamburg, Germany